Rainbows

by Grace Hansen

Abdo SKY LIGHTS Kids

Abdo Kids Jumbo is an Imprint of Abdo Kids
abdobooks.com

abdobooks.com

Published by Abdo Kids, a division of ABDO, P.O. Box 398166, Minneapolis, Minnesota 55439.

Abdo Kids Jumbo™ is a trademark and logo of Abdo Kids.

Printed in China

102019

012020

THIS BOOK CONTAINS RECYCLED MATERIALS

Photo Credits: Alamy, iStock, Science Source, Shutterstock

Production Contributors: Teddy Borth, Jennie Forsberg, Grace Hansen
Design Contributors: Dorothy Toth, Pakou Moua

Library of Congress Control Number: 2019941274

Publisher's Cataloging-in-Publication Data

Names: Hansen, Grace, author.

Title: Rainbows / by Grace Hansen

Description: Minneapolis, Minnesota : Abdo Kids, 2020 | Series: Sky lights | Includes online resources and index.

Identifiers: ISBN 9781532189104 (lib. bdg.) | ISBN 9781532189593 (ebook) | ISBN 9781098200572 (Read-to-Me ebook)

Subjects: LCSH: Rainbows--Juvenile literature. | Colors--Juvenile literature. | Refraction--Juvenile literature. | Light--Juvenile literature. | Optics--Juvenile literature. | Rain and rainfall--Juvenile literature.

Classification: DDC 535.6--dc23

Table of Contents

Illusions in the Sky

Have you ever seen a rainbow? Did you know that it wasn't really there? Rainbows are **illusions** in the sky!

Light and Energy

Light is a wave. Light is also energy. The energy of light is tied to its **wavelength**.

Red has a longer **wavelength**.

This means it is low energy.

Violet has a shorter wavelength.

This means it is high energy.

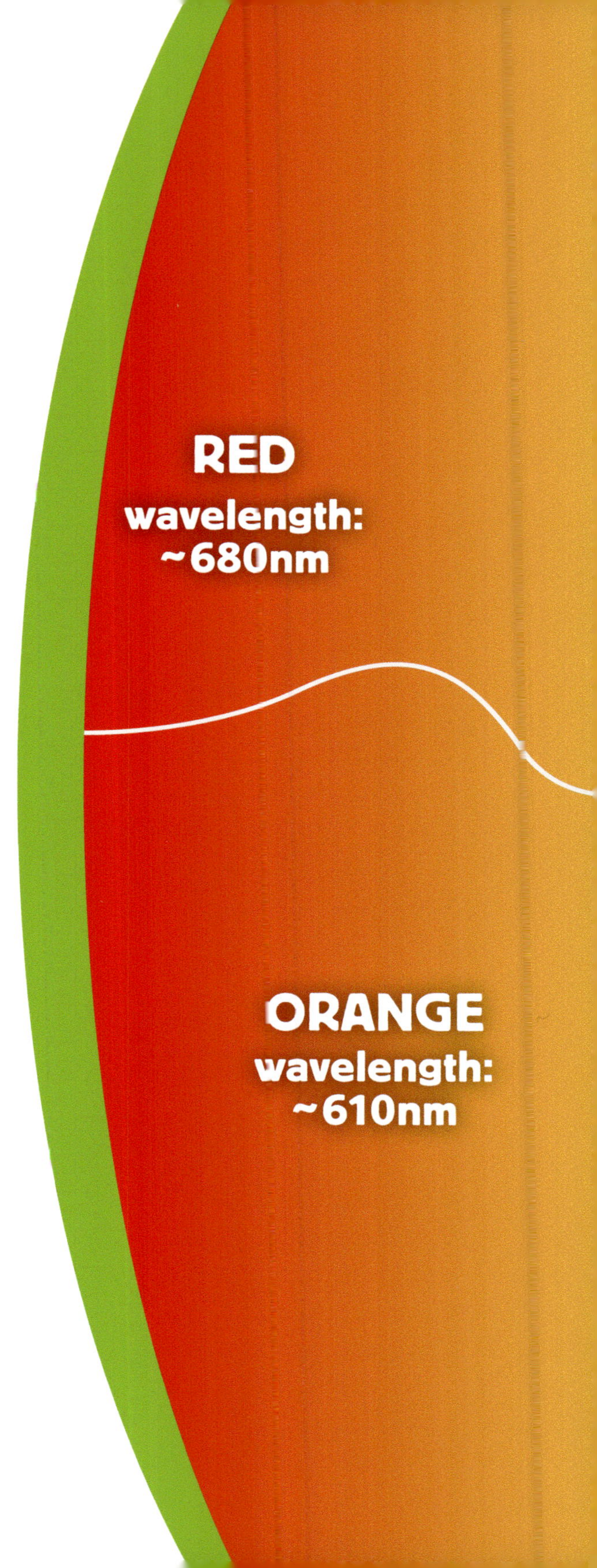

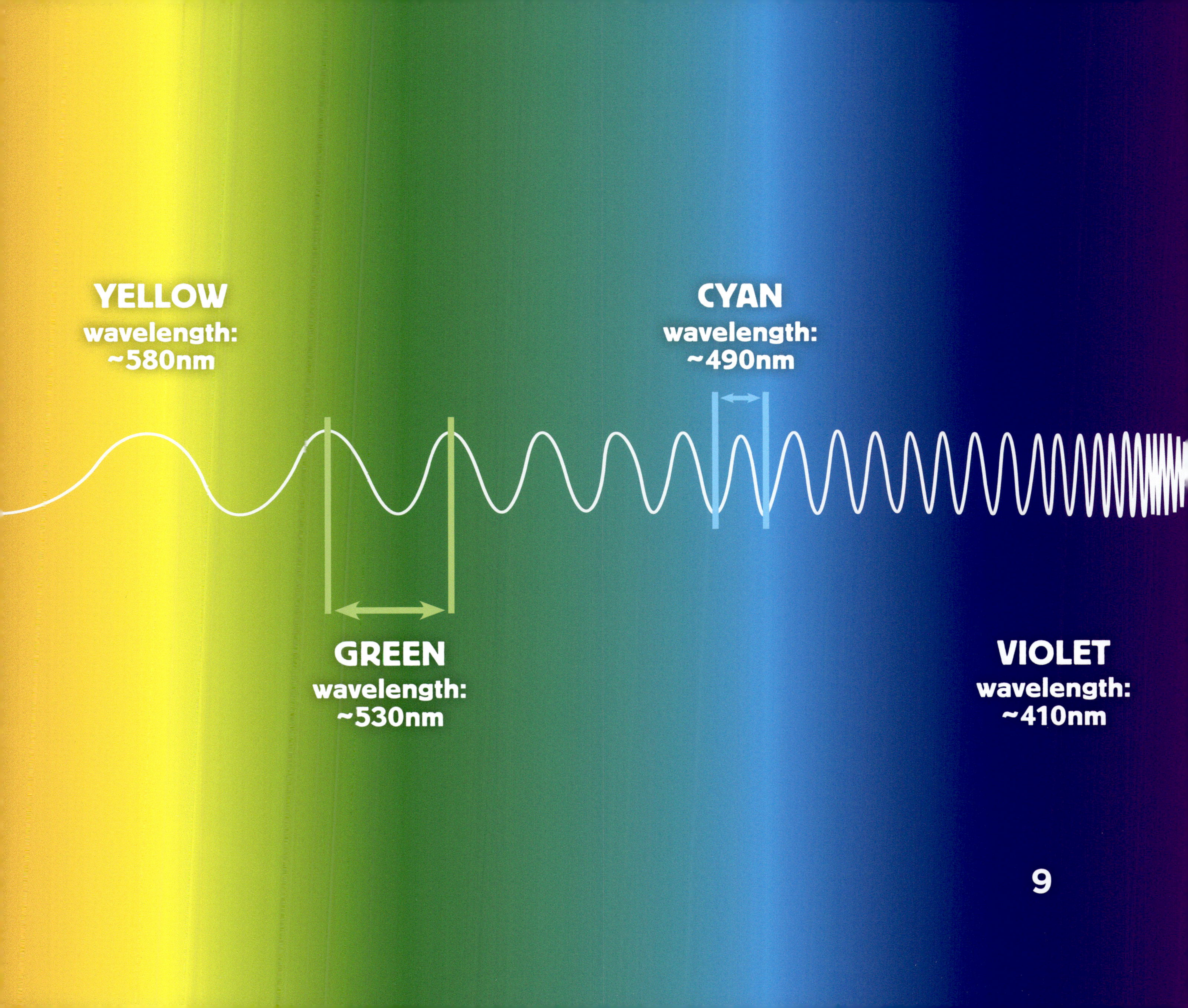
YELLOW
wavelength:
~580nm
CYAN
wavelength:
~490nm
GREEN
wavelength:
~530nm
VIOLET
wavelength:
~410nm

The sun's light waves are the only ones we can see. Our eyes see these waves as colors. All of these colors together are called white light.

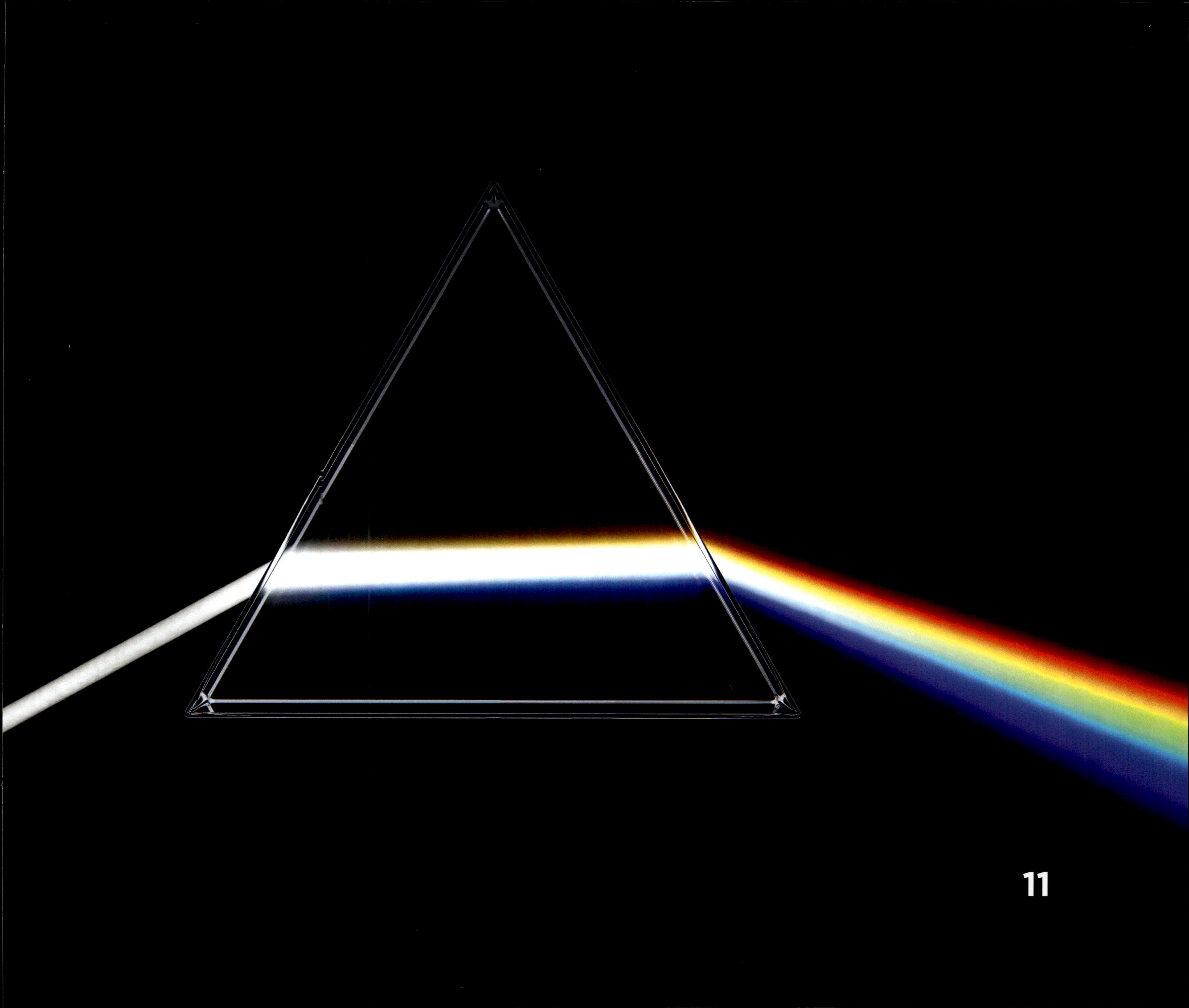

How to See a Rainbow

We often see rainbows in the early morning or late afternoon. This is because the sun is lower in the sky.

To see a rainbow, your back must be to the sun. You should be facing the rainy or **misty** sky.

Light from the sun hits the water droplets. As it passes through the droplets, it bends. Different **wavelengths** of light bend at different **angles**. This causes the white light to separate.

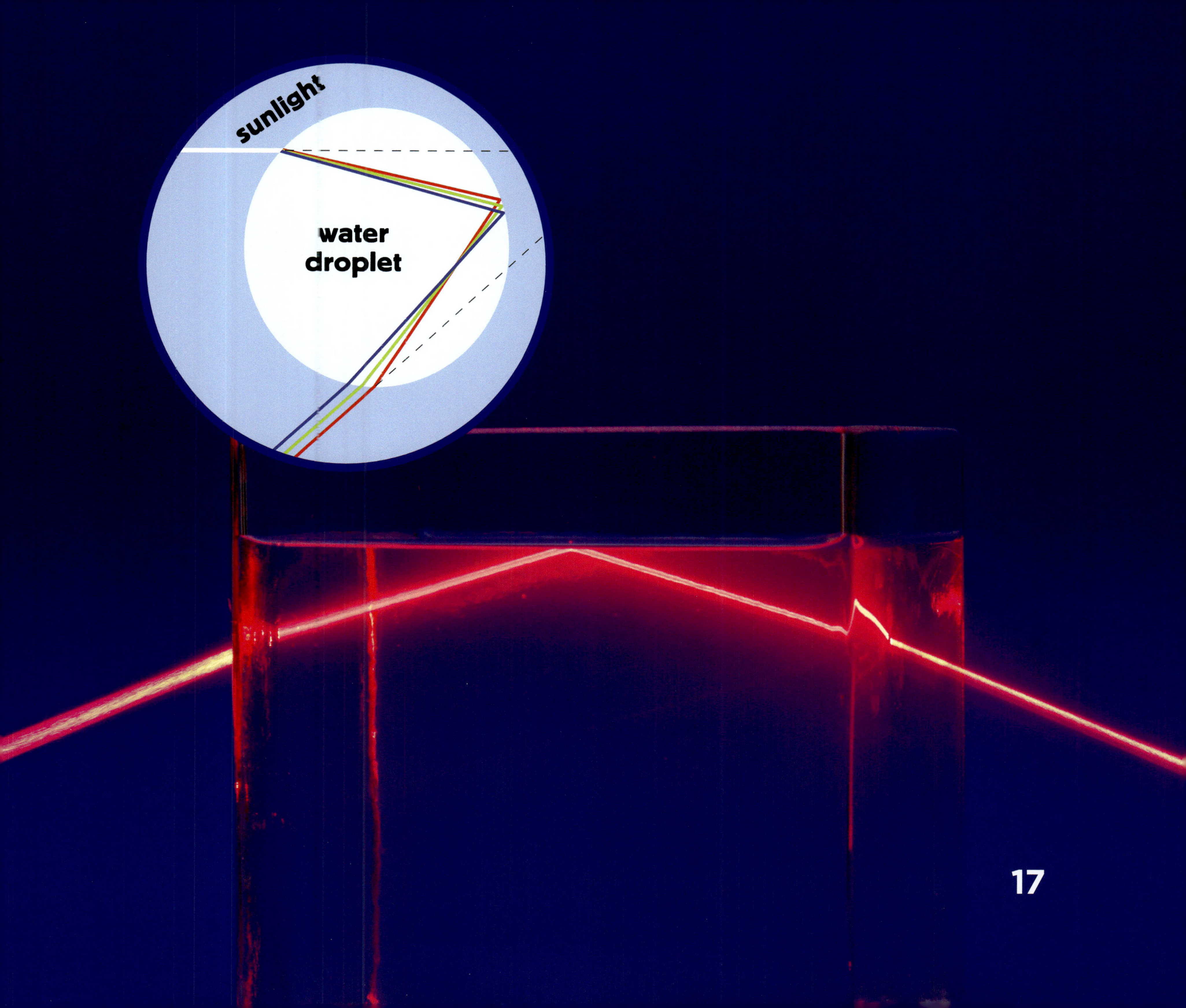
sunlight
water
droplet

The back of the raindrop then reflects the light. The light comes out of the raindrop at an **angle** back toward you. When it exits, it makes your eyes see a rainbow.

Red light makes a 42-degree **angle**. Violet's angle is 40 degrees. All the other colors fall in between. This is why the colors of the rainbow stay in the same order.

More Facts

- There are 7 colors in the rainbow. They are red, orange, yellow, green, blue, indigo, and violet.

- Even people who are standing next to each other see a different rainbow. This is because they are looking at the droplets from a slightly different **angle**.

- The lower the sun is, the higher a rainbow is in the sky.

Glossary

angle – the figure made by two lines coming from a single point.

illusion – the state of seeing things in a false way.

misty – full of mist. Mist is a cloud of tiny water droplets in the atmosphere.

wavelength – the distance between two crests of a wave.

Index